22 PRACTICAL IDEAS

WEB 2.0 TEACHER'S TOOLKIT

ELISE ABRAM B.A., B.Ed., M.Ed., O.C.T.

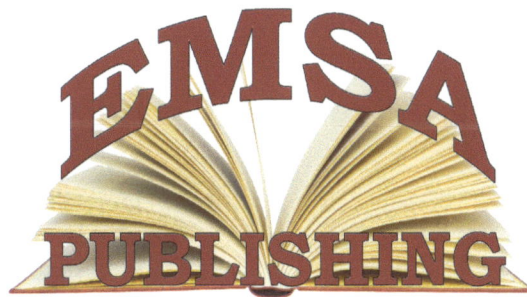

EMSA PUBLISHING

22 Practical Ideas: Web 2.0 Teacher's Toolkit

PUBLISHED BY EMSA PUBLISHING
http://emsapublishing.com

22 Practical Ideas: Web 2.0 Teacher's Toolkit is printed in Georgia with titles in Sage Sans (by Sage under the SIL Open Font License, Version 1.1.) and subtitles in Arial.

Credits:
Cover font: Sage Sans

Cover art: "Smart child—Photo" by Olly 18 under DepositPhotos Standard License. "Inspired blonde woman, tablet, social media—Photo" by Dinisis Magilov under DepositPhotos Standard License. "Grey concrete wall. Dark edges—Photo" by Stillfx under DepositPhotos Standard License.

Cover design: Elise Abram

ALSO BY ELISE ABRAM

ADULT FICTION

Chicken or Egg: A Love Story?
Revamped
Phase Shift
Throwaway Child
The Mummy Wore Combat Boots

YOUNG ADULT FICTION

Carrington Pulitzer Revelation Chronicles Online Extended Playpack
Indoctrination: The New Recruit Book Two
The New Recruit
I Was, Am, Will Be Alice
The Revenant: A YA Paranormal Thriller with Zombies

MIDDLE GRADE FICTION

Operation: Blueberry Pancakes

CHILDREN'S PICTUREBOOKS

Heddy is Sad
Harry has a lot of Energy
Luna is Afraid of Storms
Luna has Nothing to do

COMING SOON

Smiles for Africa: Adamma's Story
Things Go Bump All the Time: Wendigo
Things Go Bump All the Time: Fallen Angel

TABLE OF CONTENTS

WHY THIS BOOK?

I wrote a book in the midst of the pandemic about my best practices for online teaching. I was doing a lot of experimenting with online tools to use with my students, and I decided to record what I'd learned. When I was nearly done, we'd crested COVID's third wave, went into yet another lockdown, and I was too busy trying to cope to complete the book.

Flash forward a few months, and people were getting vaccinated. It seemed certain we would soon return to face-to-face classes full-time. I held out hope I could continue making use of interactive online applications in my daily routine, but it soon became clear there wasn't enough technology in the school to do that. But I had all this information, and I wanted to do something with it, and there was the nearly finished best practice manuscript, so I decided to finish the book with a slight shift.

When I first started teaching, I created web pages as tutorials for my students using simple HTML, but it wasn't enough. I wanted them to do more than read web page tutorials. I wanted them to be hand work in work digitally, as well. They could have emailed the assignments to me, but few students had personal emails, and board-issued emails were still many years away. To remedy this, I created forms with submit buttons. Students could paste their work into text fields and hand it in that way. It worked, but it still wasn't enough.

The Internet finally caught up to my needs about a decade or so ago with the implementation of applications dubbed "Web 2.0" (Web two-point-oh), applications that "allow users to participate directly in the creation, refinement and distribution of shared content"(Selwyn). This pairs well with the shift in learning philosophies andchange in focus over the past few decades from teacher as the font of all knowledge to facilitator of student-centred learning. This has

> led to the emergence of...technology enhanced learning (TEL) including: (i) a shift from a focus on content to communication, (ii) a shift from a passive to a more interactive engagement of students in the educational process, and (iii) a shift from a focus on individual learners to more socially situated learning (Conole, 2007), (Rahimi, et al.

40)

to support a "student-centric" (Rahimi, et al. 45) and "more activity-oriented" (Rahimi, et al. 45) approach, rather than learning that is "lecture based" (Rahimi, et al. 45). This is important because not only does it help "develop...knowledge, but also... skills and resources which are equally necessary to bring a social and technological change resulting into a continuous lifelong learning" (Ozcinar, et al.), something teachers have been attempting to foster in their students for years.

This book of best practices is important when one considers that studies suggest that most of the teachers do not give enough room for technology-supported content in their lessons (Ciftci, 2013), are not sufficiently ready to utilize [the] internet [sic] and computer for teaching purposes (Erdemir, Bakirci & Eyduran, 2009; Hsu, 2016), believe in the importance of technology in education but feel incompetent to use it in their fields (Duhaney, 2012; Hirça & Simsek, 2013), [see] pedagogical beliefs as a barrier in technology integration (Tondeur, van Braak, Ertmer & Ottenbreit-Leftwich, 2017) and [this is why] many teachers do not use technology in education (Can & Kaymakci, 2016). (Tatli, et al 2)

It is the hope that this book will help teachers who feel they aren't ready to utilize the Internet and computers in their practices or who are insecure in their knowledge when it comes to technology integration. Hopefully, this book will help these teachers to see how easy and user-friendly Web 2.0 apps really are. Hopefully, this book will give teachers who are already comfortable with technology-based alternate ideas as to how they might use these apps in their practice.

The most insurmountable barrier to implementing Web 2.0 apps in the classroom as I see it is not teacher know-how, willingness, or ingenuity but the availability of computers and/or Chromebooks in their schools school. In that case, it is my sincere hope that they will be able to navigate the uphill battle of finding and/or booking technology in their schools so they can implement some of these ideas because they work really well for student engagement and accountability. They are also really fun to consider when planning.

OVERVIEW

Though many teachers dabbled in Web 2.0 tools prior to the pandemic and "emergency online learning," they were forced to indulge once classes were forced online. In Ontario, many teachers were left scrambling as we were often given 24 hours notice prior to the shift online, and we had little to no training in using online tools to foster student engagement and accountability. Even though many teachers were already versed in *Google Classroom*, few of us used the full suite of *Google* apps, and that had to change. Needless to say, there was a steep learning curve. Add to that the fact that many school boards maintain a list of available apps, categorizing them into those you can automatically use, those you may only use with parental consent, and those you should not use at all due to security issues, and teachers are limited as to which apps they may use while teaching.

Another obstacle is that technology is not always readily accessible. During the lockdown and stay-at-home learning, many school boards ensured students had access to technology. With the return to class, school boards have repossessed the technology, leaving teachers scrambling to find adequate tech in numbers plentiful enough to support students who do not have it. Still, with planning and forethought, most teachers are able to book cross-curricular labs to assist their utilization of Web 2.0 tools in their classroom.

This book is an overview of Web 2.0, what it is, and how it might be employed in a high school (primarily English language and literature) classroom.

WHAT IS WEB 2.0?

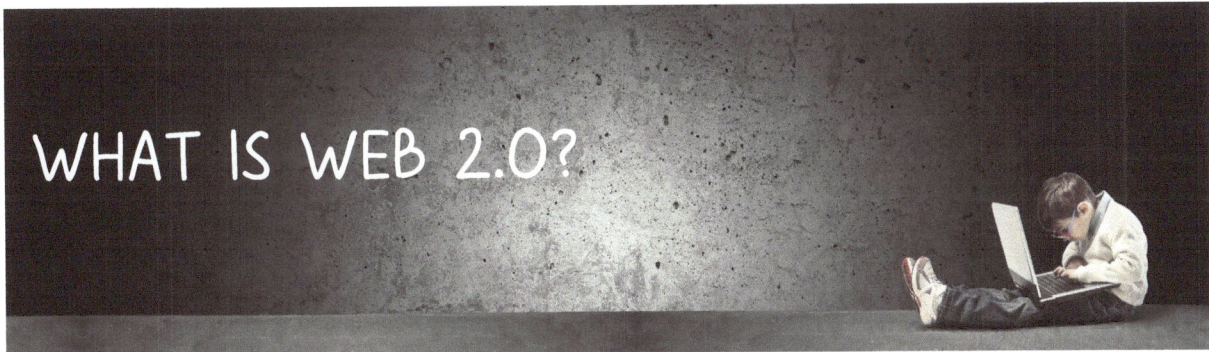

The moniker Web 2.0 was coined to indicate all of the interactive tools available online. These tools

> enable users to create, share, collaborate and communicate their work with others, without any need of any web design or publishing skills. These capabilities were not present in [the] Web 1.0 environment, (Lipika)

characterized by static, read-only web pages. The key to Web 2.0 apps is that

> [t]hey are interactive, multi-purpose, easy-to-use digital platforms that encourage students to collaborate with each other or create and share individualized response products [in] engaging ways students can interact with, and most importantly, learn from course material. (Rowe and Chapel)

It is, however, important that

> [t]he tool must actually enhance the learning process, not simply add unnecessary tasks for students to complete. If students can communicate their understanding of the learning objectives without technology, then a more traditional response assignment may suffice. (Gulley and Thomas)

Having said that, when possible, I like to use technology as it seems to engage students more than traditional pen and paper responses.

The "benefits of using Web 2.0 in teaching include" (An, et al. 1):
- "interaction, communication, and collaboration" (An, et al. 1);
- "ease of use and flexibility" (An, et al. 1);
- "writing and technology skills" (An, et al. 1);
- "build[ing] a sense of community" (An, et al. 3);
- not limiting students to "their own artistic abilities" (Gulley and Thomas);
- being "available at any time, any place" (Lipika);
- actively involving "learners... in knowledge building" (Lipika);
- tracking "every edit that has been made" (Lipika);
- "quick and easy access to all kinds of information and content" (Tatli, et al. 3);

- students "becom[ing]...contributor[s] to the Internet...giv[ing] [them] the opportunity to create" (Zappa); and
- "provid[ing] engaging ways students can interact with, and most importantly, learn from course material" (Gulley and Thomas).

Dubbed personal learning environments (PLEs) by McLoughlin & Lee (2010),
> [their] conceived goal...is to enable students, not only to consume content, but to remix, produce, and express their personal presentation of knowledge. Furthermore, it has been argued that PLEs presume and support an active role for students by placing them in the center [sic] of their learning processes, corroborating their sense of ownership of learning, and enhancing their control in educational process[es]. (Downes, 2006; Buchem, 2012) (Rahimi, et al. 56)

In my practice, I frequently assign research essays in which students take to the library and/or online to search for "expert" quotations to support their opinions. Rahimi, van den Berg, and Veen, in their article "A Pedagogy-driven Framework for Integrating Web 2.0 Tools into Educational Practices and Building Personal Learning Environments" describes this process as
> the appropriation of content by students. Appropriation [is] the "ability to meaningfully sample and remix media content" (Jenkins, 2006) makes student simultaneously... the producer and consumer of content and can be understood as a learning process in which students learn through picking several content (sampling) and putting them back together (remixing) to produce new content and knowledge objects such as ideas, discussions, conversations, comments, replies, concept maps, webpages, podcasts, wikis, and blog posts (Jenkins, 2006). In addition to essays, this really describes most assignments [with] research components.

Further, the Ontario curriculum document for English, *The Ontario Curriculum Grades 9 and 10: English*, includes expectations such as:
- "determine whether the ideas and information gathered are relevant to the topic, accurate, and complete and appropriately meet the requirements of the writing task" (i.e., finding appropriate research to "appropriate" and incorporate into their writing);
- "use a variety of presentation features, including print and script, fonts, graphics, and layout, to improve the clarity and coherence of their work and to heighten its appeal for their audience" (i.e., across multiple platforms);
- "produce pieces of published work to meet criteria identified by the teacher, based

on the curriculum expectations" to change-up the traditional paper and pencil format of most school assignments;

- "identify general and specific characteristics of a variety of media forms and explain how they shape content and create meaning," which is exemplified when utilizing a variety of platforms for a number of assignments;
- "identify a variety of conventions and/or techniques appropriate to a media form they plan to use, and explain how these will help them communicate specific aspects of their intended meaning" if students are given a choice as to platform; and
- "produce media texts for a variety of purposes and audiences, using appropriate forms, conventions, and techniques," justifying choiced made when completing projects incorporating Web 2.0 tools,

In short, Web 2.0 apps, by their very nature,

can enrich the learning experiences of students and nurture their cognitive skills by providing them opportunities to practice "learning by doing" (Brown, Collins, & Duguid, 1989)... experience "learning with technology" (Jonassen & Reeves, 1995), and construct...personal presentation[s] of knowledge and share [them] with others. (Rahimi, et al. 48)

In this way, the teacher fosters "an active 'architecture of participation' rather than site of passive consumption" (Selwyn).

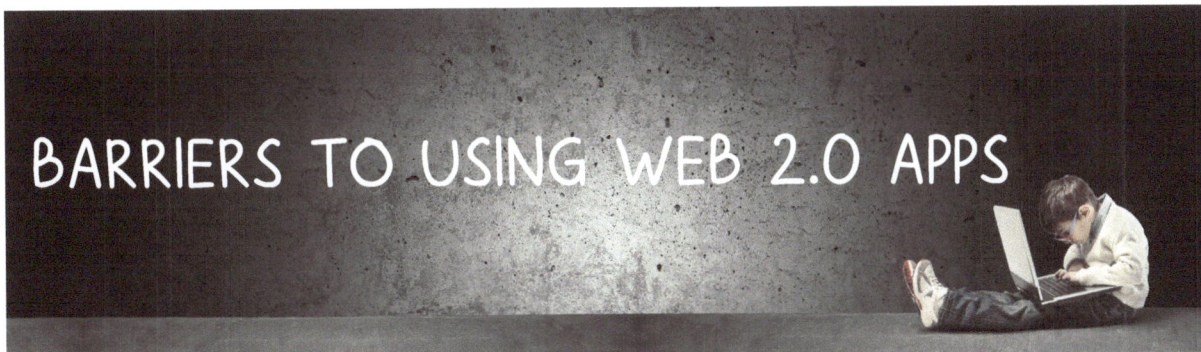

BARRIERS TO USING WEB 2.0 APPS

LACK OF TECHNOLOGY

As previously mentioned, technology or the lack thereof is, perhaps, the largest barrier to implementing the regular use of Web 2.0 tools in the classroom. As I write this, I am teaching blended learning in a school comparatively rich in technology that is difficult to access. As many as half of my students (in a class of 31) do not have access to technology beyond their phones (with the exception of the few students online). While phones are good for writing short blurbs of text over a short period of time, I wouldn't want to write an entire essay on one, nor would I expect my students to. It is, therefore, necessary to provide technology for those without it.

I teach in a school where I have access to three portable computer or Chromebook labs, with about 26 devices in each. Typically, I take five or six computers from one of the labs to supplement my class, but in one class, I have to use at least half of the computers on the cart. Luckily, the room in which I teach has a cart in it, but the cart is subject to booking, so I never know when I will be left with no tech—or not enough tech—to supplement the learners in need. I can book the cart myself, but it is generally frowned upon to book it for more than three days in a row.

STUDENTS MAY NOT BE TECH-SAVVY

No matter how much teachers know about technology, their students will know more...or so I've been told at practically every technology workshop I've ever been to, but that's far from the truth. Most students can't figure out how to use *TurnItIn* or *Google Classroom*, even with step-by-step instructions, let alone other more complicated apps like *Google Sites* and other learning management systems (LMS).

TECH IS OUT OF DATE

Even when I am lucky enough to have full access to a computer cart, the technology is old. Many of the computers can take up to 15 minutes to boot and more to log in, and most of the batteries don't hold a charge for the full 150 minutes of class. We share the carts, so while I might take 15, other teachers are free to come in and take as many as they need, so if the battery in one of my student's computers dies, chances are there won't be one available to replace it. I have been offering to print handouts for the students when this happens, but most opt to work on their phones or opt out of the activity altogether.

TEACHER MONITORING

When students use technology in the class (be it school- or student-owned), it is near impossible to monitor what they are doing on it. In the days of sedentary computer labs, most computers faced the centre of the room, making it easy for teachers to see the monitors. There was also sotware installed for teachers to track student activity on a desktop computer at the teacher's desk, but with laptops and phones being used in the classroom, this is no longer the case. Teachers have to trust that the students are doing their work, but students, being kids, don't have that kind of integrity in grades nine or ten (and sometimes not even in grade 12), which has the potential for disaster and far from sets students up for success.

WEB 2.0 IN THE CLASSROOM

Even though there are many barriers to using interactive tools in the classroom, the positives far outweigh the negatives. What follows is a series of best practice ideas to give the reader an idea of what is possible, but this barely scratches the surface. Given the number of Web 2.0 tools available and more being developed on a daily basis, it would be virtually impossible to pen a comprehensive, all-encompassing guide. Here is what I know so far; I look forward to seeing how this changes and grows in the future.

1. DIGITAL WORKSHEETS

I first realized the potential of digital worksheets when I clicked on a *Facebook* ad and downloaded a sample. Since then, I have discovered myriad sites online describing what they are and how to use them in a number of subjects.

Interactive digital worksheets are similar to paper and pencil worksheets except they are done in a Web 2.0 tool like *Google Slides*. There is space left for students to fill in the blanks, and they sometimes have moving parts (like things you can drag and drop to answer questions).

These take a bit of preparation, as the pieces that students drag and drop must first be made into graphics. To do this:
1. Take a screenshot of your worksheet.
2. Cut it up into a series of graphics using software.
3. Reconstruct the worksheet in *Google Slides*.

At its most basic, a digital worksheet can be a series of questions on a series of slides. In the worksheet in Figure 1, students simply click beside the first bullet point and type in their answers. They click below the caption "Add a note to explain what these symbols mean" and begin typing. Next, they grab a push-pin and move it mark the symbols. In other examples, they might use a star to "vote" on an option. Prior to sharing with students, the teacher selects a graphic to use and "stacks" copies of the graphic one on top of the next, giving the appearance of a single push-pin when there are actually multiples. In this case, the teacher can estimate the number of symbols and include enough push-pin graphics plus one or two extras for students to use. The distribution of push-pins is the "moving part" alluded to earlier. Note that the slide in Figure 1 was a part of a larger slide deck composed of a series of notes and digital worksheets, taking students through the process of analyzing satirical texts (including cartoons, writing, and videos).

Figure 1: Digital worksheet using Google Slides. Image by Markusszy.

This works well because you can check students' work at a glance, given the layout of the slides.

Another app that works well for creating digital worksheets is *Jamboard*, a digital whiteboard app that is a part of the *Google* suite of apps. In addition to patterned and coloured backgrounds, *Jamboard* allows you to upload an image as the background of a slide.

In Figure 2, I took a screen capture of a black line master (BLM) of a Freytag pyramid and used it as the slide background. The background is composed of the title "Freytag Pyramid," an image of a Freytag pyramid, and a space to record story title and author's name.

The activity in Figure 2 was created as a way to take gauge students' knowledge of elements of a short story. I projected this slide onto a screen at the front of the class and filled it ou

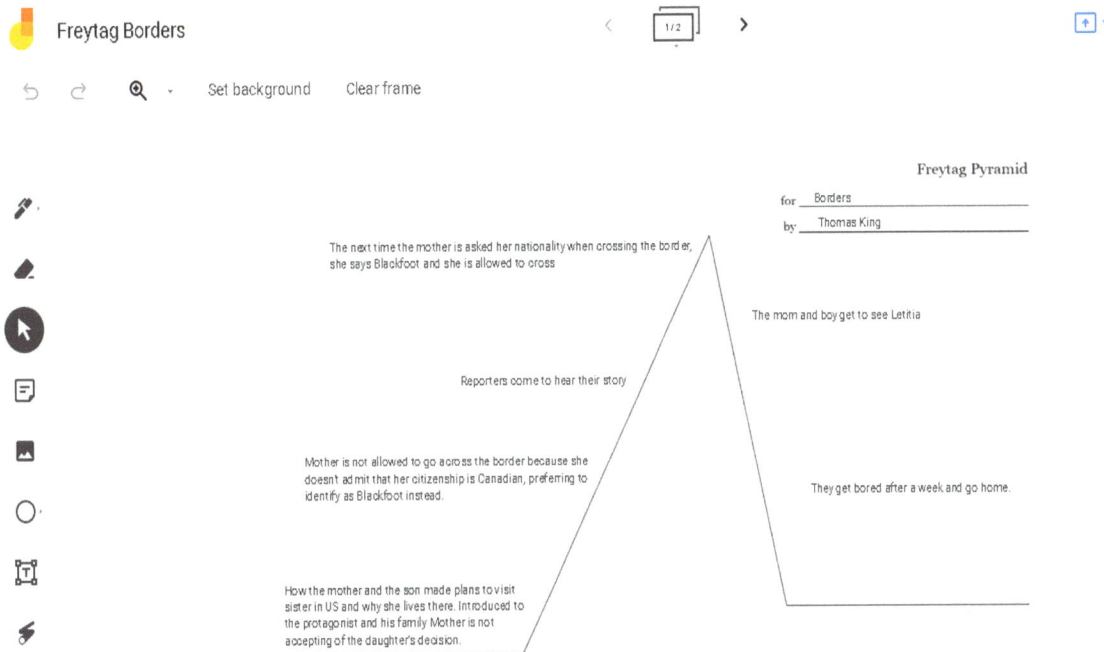

Figure 2: Digital worksheet of Freytag pyramid.

as the students sugggested answers. Students were then given "view" status so they could use this as an exemplar moving forward. Alternate modes of use include (but are not limited to):

- sharing the *Jamboard* with students who were given editing status and asked to do one slide collaboratively or
- sharing the background image and *Jamboard* with students, asking them to create their own slide, import the image, and complete their own Freytag pyramid. In this case, the background image could be modified to include a space for students to put their own names alongside the story title and author.

To track participation in *Jamboard*, click on the three stacked dots beside the blue "Share" button and choose "See version history." Alternately, you can "take attendance" of who has logged into the app while the activity is ongoing.

Figure 3 is another example of a digital worksheet in Jamboard. The goal of this activity is to demonstrate how to brainstorm an essay topic to derive claims and a thesis. The worksheet itself is only the idea web, which is the same as a BLM I had given students the day before. I posted the essay topic ("Question") in the centre prior to introducing the activity to students. Students suggested essay topics, each of which I transcribed and posted on the idea web. When we were done brainstorming ideas, I reminded students

Web 2.0 Teacher's Toolkit 19

of how to write a thesis statement (Salazar) and typed that in at the top right of the slide.

During the process, it became apparant that one of the ideas brainstormed would make a better thesis than a claim, so I copied it into the top left corner of the slide and with the students' help, massaged it into a three-part thesis, which I wrote at the bottom left of the slide. When we were done, I shared the slide with students so they had an exemplar of the task they were about to complete.

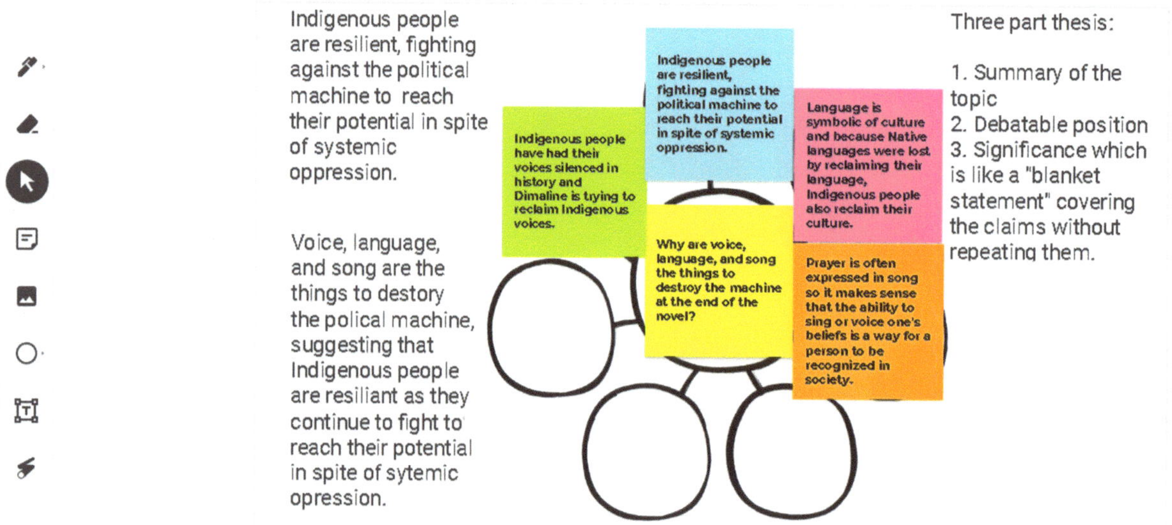

Figure 3: Screen capture of digital worksheet in *Jamboard*.

Google Forms is another tool you can use to create digital worksheets.Here are some examples from a worksheet on theme I used recently. The worksheet is in three parts to reflect the current preferred structure of lesson plans:

Part 1: Minds On
This section introduces the lesson, "getting students' minds focused on the topic of the lesson" (Cooke). Though I chose to do a diagnostic task for this lesson, teachers may assign students to watch a video, read an article, do a placemat, complete a critical inquiry task, or have a discussion with an elbow partner or group, to name a few.

Part 2: Action
The "Action" part of a lesson helps students "to make meaning of the topic or skills in the lesson." According to the Toronto Catholic District School Board Prezi on "Minds On, Action, Consolidate" (Cooke), the purpose of this part of the lesson is to "explore and

investigate (and struggle) with a new concept prior to being told" (Cooke). Further, it allows for students to navigate the topic together, "support[ing] each other in their learning" while "construct[ing] new knowledge...develop[ing] concepts using higher order thinking skills" (Cooke) as well as "allow[ing] the teacher to identify and challenge student misconceptions" (Cooke) of the topic. This section of my worksheet questioned students as to possible themes of a story; differentiating between theme, symbolism, character, plot, and other story elements; and reinforcing the format of a thematic statement (Boyd). Lastly, it asks students to write a thematic statement for the current text being studied and takes them through an evaluation process to self-assess if their thematic statement is a good one.

Part 3: Consolidation

The "Consolidation" of the lesson "pull[s] together [the] learning for the day related to the learning goals" (Cooke). In this particular lesson, students were asked to come up with an alternate thematic statement using thematic topics of the students' choosing. Alternately, the teacher may brainstorm a list of thematic topics with students from which they may choose. The consolidation begins with where they ended the action phase, creating a thematic statement, but it goes one step further, asking students to find a relevant quotation and add an explanation connecting their quotation to their thematic statement to discuss that particular theme in the text.

MINDS ON:

How much do you know about theme in literature?

Theme is defined as the main idea or underlying meaning a writer explores in a 1 point
novel, short story, or other literary work. *

○ true

○ false

The theme of a story can be conveyed using characters, setting, dialogue, 1 point
plot, or a combination of all of these elements. *

○ true

○ false

Theme is a summary of the story. * 1 point

○ true

○ false

Figure 4: Sample of "Minds On" diagnostic questions for Theme worksheet using Google Forms.

Action

Complete each of the following questions to the best of your ability.

Which of the following might be the theme of a story? *   1 point

○ A 35-year-old woman named Essie and her brother

○ It's about being honest and reveals that telling the truth may cause pain, but in the end, it's better than lying.

○ Essie lied to her brother about her identitiy for two years, but she finally decided to tell him the truth.

○ a small apartment in Marfa, Texas

Which of the following might be the theme of a story? *   1 point

○ It's about facing one's fears and reveals fear can be more dangerous than any beast

○ a fear of heights

○ 'I'll get you, my pretty, and your little dogs, too!'

○ A brave young girl pretends to be a man and takes her father's place in the army.

Figure 5: Sample of "Action" questions for Theme worksheet using Google Forms.

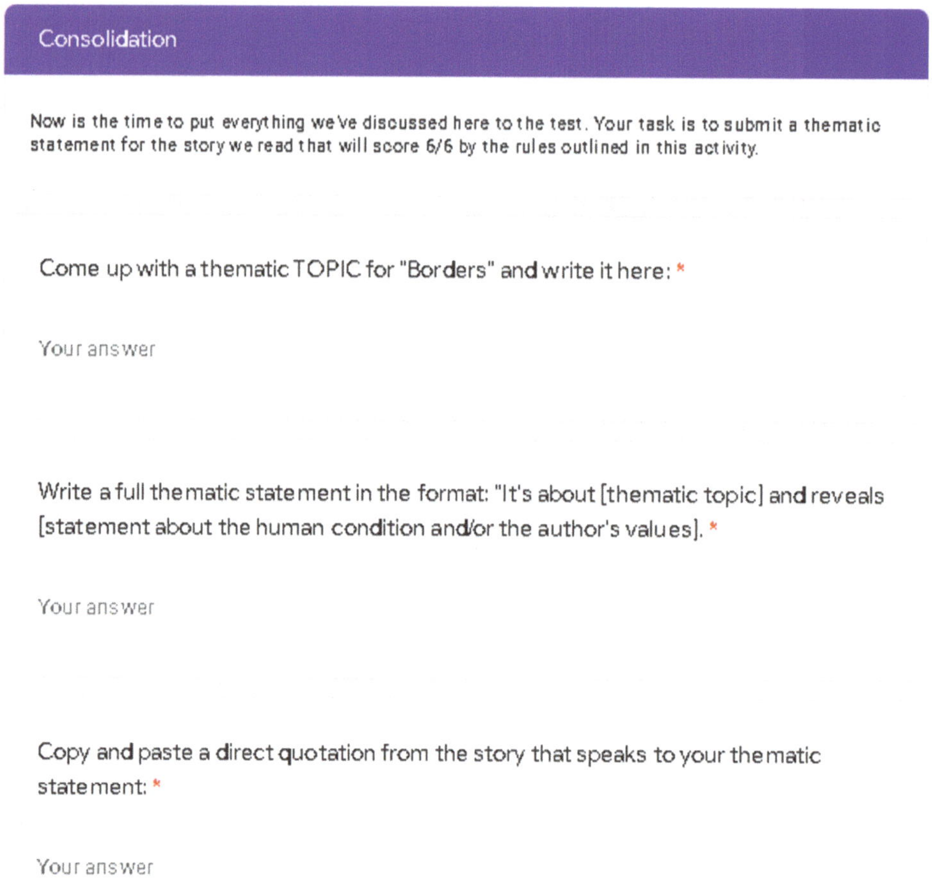

Consolidation

Now is the time to put everything we've discussed here to the test. Your task is to submit a thematic statement for the story we read that will score 6/6 by the rules outlined in this activity.

Come up with a thematic TOPIC for "Borders" and write it here: *

Your answer

Write a full thematic statement in the format: "It's about [thematic topic] and reveals [statement about the human condition and/or the author's values]. *

Your answer

Copy and paste a direct quotation from the story that speaks to your thematic statement: *

Your answer

Figure 6: Sample of "Consolidation" questions for Theme worksheet using *Google Forms*.

2. CRITICAL CHALLENGES

Once completed, this activity leads to a critical challenge in which students are given a number of choices and asked to order the answers from most important to least, or choose which item doesn't belong, or decide which is the most suitable, and so on. Not only are students asked to complete the challenge (which should engage critical thinking and/ or inquiry), but they must be prepared to justify their answers (James). Examples from students' work is kept anonymous so (hopefully) no one seems slighted by the criticism.

To set up a critical challenge, create a spreadsheet of student answers and select a number of student-generated thematic statements culled (by the teacher) from the consolidation portion of the worksheet. The thematic statements selected should be of various levels, from correct to not a thematic statement at all. Examples from students' work are kept anonymous, so (hopefully) no one is slighted by the criticism.

Lay these choices out in a *Word* or *Google Docs* document and take a screenshot. Use the image generated as the background of a *Jamboard* slide per Figure 7.

Review instructions with students as displayed on the slide and give students time to complete the activity. Note that the original sticky pages had the students' complete names on them so I could call on them to ask why they had marked that particular thematic statement as either good or bad. This activity generates discussion as to the parameters of a good thematic statement. It can also lead to an exit ticket activity in which students are asked to select one of the down-voted thematic statements and edit it to make it better so you can gauge the effectiveness of the overall lesson.

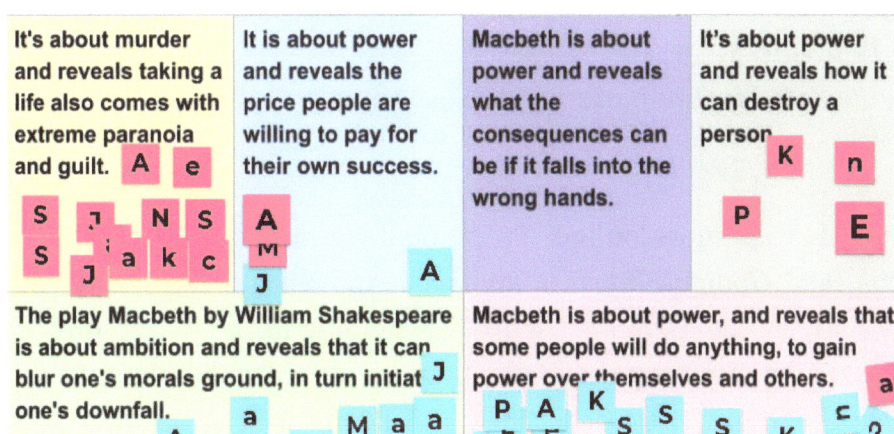

Figure 7: Screen capture of critical challenge in *Jamboard*.

Another example might be to allow students to post their opinions as to why Shakespeare is still taught in school today (this could also be done in a poll in advance, and the critical challenge could take place during the next class). Students are given their peers' thoughts and asked to pick the most and least relevant (again, by "voting" with coloured stickies), and this is used to spark conversation.

A special thanks to Usha James from the *Critical Thinking Consortium* for introducing me to the idea of critical challenges.

The possibilities for worksheets are endless, and they can be as simple or complicated as you dare make them.

3. EXIT TICKETS

According to *Edutopia*

> Exit tickets [also called exit cards] are a formative assessment tool that give teachers a way to assess how well students understand the material they are learning in class...A good exit ticket can tell whether students have a superficial or in-depth understanding of the material. ("Gaining Understanding")

They can be a great way to take the temperature of a class, poll levels of student engagement, collect information, and get students to ask questions they might not ordinarily ask. Sometimes, exit tickets can also let you know that a question you asked wasn't well-phrased when you don't get the types of answers you expected.

Some sites providing excellent examples of fun exit-ticket ideas include:
- *The Owl Teacher: 24 Exit Ticket Ideas* (Tammy),
- *Sample Exit Tickets* (Wakeford),
- *Exit Slip Templates,*
- *24 Printable Exit Slip Templates*, and
- *Exit Slips.*

There are literally tons more online. Traditionally administered via paper and pencil, *Google Forms* (and *Google Slides*) is an excellent Web 2.0 tool with which to assign exit tickets for student completion. Figure 8 is notable for a few reasons. First, it has a fun graphic at the top, inviting the students in by providing interest, as does the three questions from the original exit ticket idea (see Tammy). It also works well if there is a word count suggested to ensure students use depth of critical thought and analysis in their answers. Students are told:
- no length specified = a sentence
- ca. 50 words = a few sentences
- ca. 100 or 200 words = a paragraph

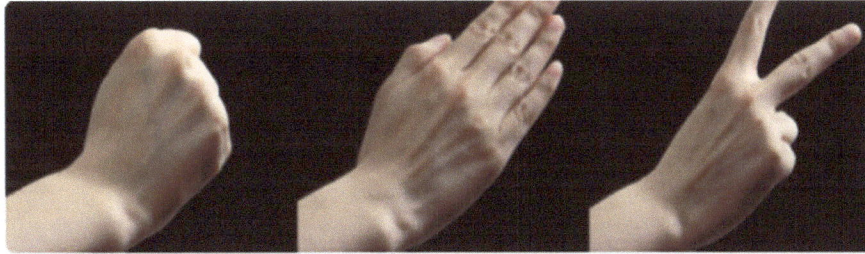

Rock, Paper, Scissors Exit Ticket

Answer each of the following questions regarding your learning so far this semester. Pay attention to the word count. Circa (about) 50 words means your answer should have about 2-3 sentences. Use complete sentences with proper English spelling and grammar in your answers.

elise.abram@gappa.yrdab.ca Switch account ☁

Your email will be recorded when you submit this form

* Required

First and last name *

Your answer

What do you feel was the 'rock' of the lesson (the solid, foundational part)? (ca. 50 words) *

Your answer

What do you feel should definitely be written down (on paper) to remember? (ca. 50 words) *

Your answer

Figure 8: Rock, Paper, Scissors exit ticket after Tammy.

With *Google Forms*, you can see student answers at a glance under the "Summary" tab, letting you know how many students answered each answer correctly:

🔊 Insights

Average	Median	Range
20.46 / 24 points	21 / 24 points	13 - 24 points

Total points distribution

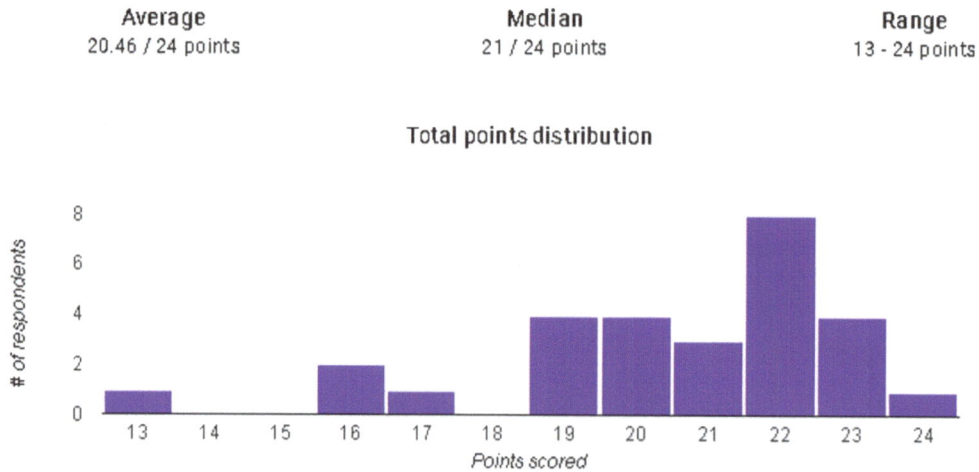

Figure 9: Question by question breakdown of correct answers.

Theme is defined as the main idea or underlying meaning a writer explores in a novel, short story, or other literary work.

27 / 28 correct responses

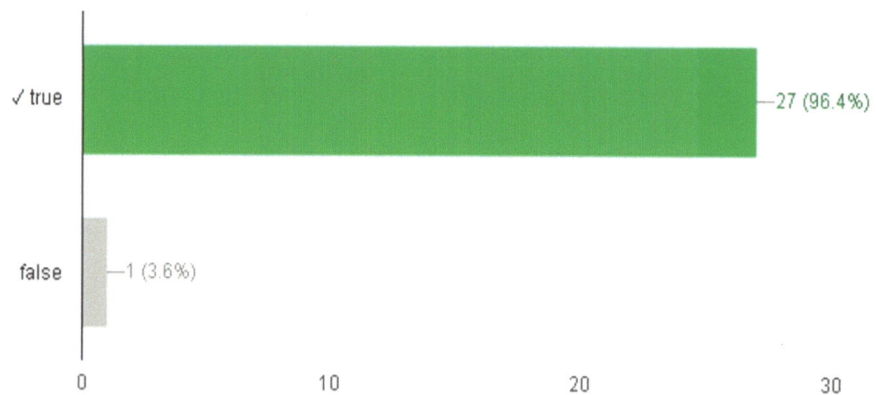

Figure 10: Individual question statistics.

Given that exit tickets are an excellent way to poll students to "help you assess if students have 'caught what you taught' and plan for the next lesson or unit of instruction" (Randall, et al.), *Google Forms* are an excellent way to accomplish this task.

4. MULTIMEDIA ASSIGNMENTS

In Ontario, the English curriculum document mentions multimedia as an "[i]nformation and communications technologies (ICT)" tool to "significantly extend and enrich teachers' instructional strategies and support students' language learning." It goes on to say that "ICT tools include multimedia resources" (35). Later, it says, "[w]henever appropriate...students should be encouraged to use ICT to support and communicate their learning" (35). Although there are not specific expectations mentioning multimedia, there is a host regarding creating media texts, regarding purpose and audience, form, conventions and techniques, and producing "media texts for several different purposes and audiences, using forms, conventions, and techniques" (53) in every course. Due to the nature of Web 2.0 tools, they make the perfect platform for creating multimedia assignments to meet many expectations, including those relating to media studies.

Caveat #1
There is a general expectation that teachers will not assign the use of a piece of software for assignment completion without first teaching its use. It is therefore imperative that, prior to assigning a multimedia assignment (be it on *Jamboard*, *Padlet* (see Figure 11), *Google Sites*, *Google Slides*, and the like), teachers use apps in interactive tasks with students earlier in the semester to build familiarity, skill, and maybe even mastery.

Caveat #2
The second expectation is that teachers not give assignments (be it diagnostic, formative, or summative) requiring the use of Web 2.0 apps without first providing students with the appropriate technology to set them up for success. While it might be okay to expect students to complete simple *Google* forms or play a game of *Kahoot!* on their phones, this is not acceptable for larger, longer assignments.

That being said, there are a number of Web 2.0 tools that may be used as platforms for creating multimedia assignments. Though some may have their limitations—for example, not allowing for the embedding of video and audio as *Google Slides* does—they, nevertheless, lend themselves to the creation of rich, multimedia assignments.

Jamboard allows students to create multiple whiteboard "pages" in a single file.

Some elements students (and teachers) can incorporate in a multimedia assignment using *Jamboard*:

- colour
- images
- backgrounds
- text with a number of font faces and sizes
- sticky-notes (for added interest in displaying text)
- lines and shapes
- important quotations and commentary
- adding URLs (instead of embedded links)

Figures 11, 11a, and 11b are examples of multimedia assignment utilizing Jamboard.

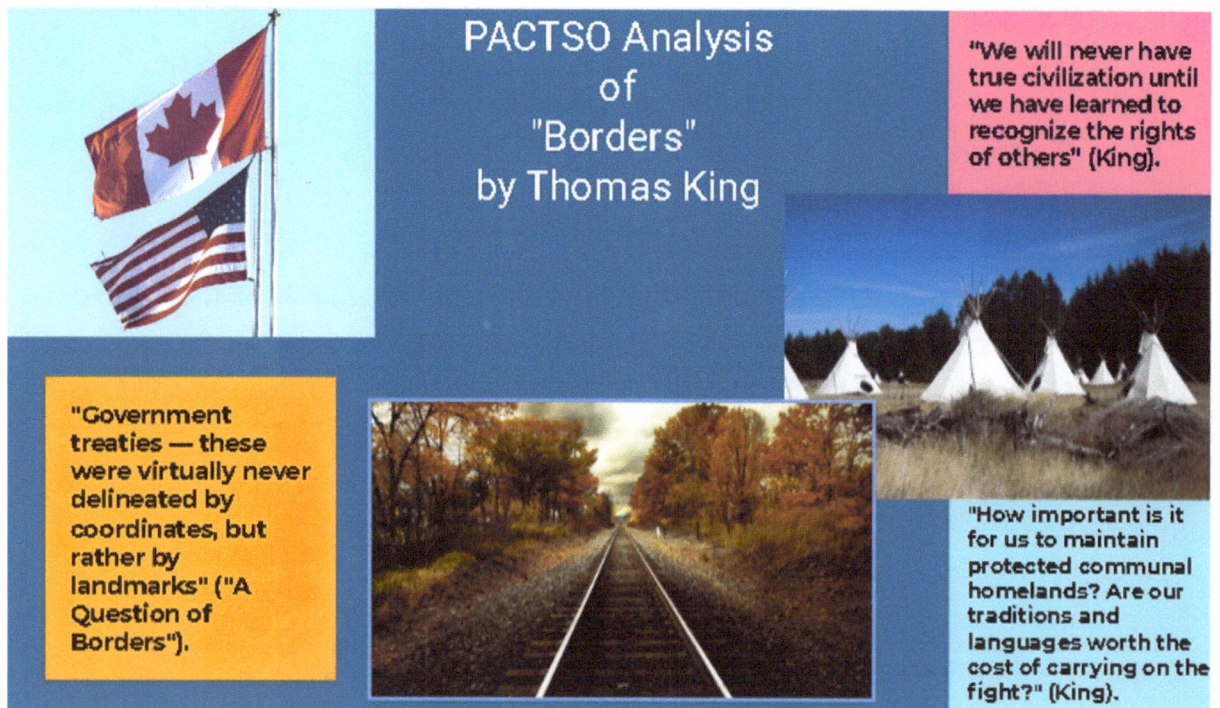

Figure 11: Two slides of a *Jamboard* multimedia presentation.

TEXT-TO-WORLD CONNECTION

There is a text-to-world connection to be made when the central conflict in "Borders" is the boy's mother refusing to claim her Canadian or American citizenship. The first time this happens, the customs guard says, "I know that we got Blackfeet on the American side and the Canadians got Blackfeet on their side. Just so we can keep our records straight, what side do you come from?" (135). Borders across countries are an arbitrary European invention, and the fight over whether or not Indigenous rights transcend the border continues today. In a CBC article dated 28 December 2018, "Marc Miller, parliamentary secretary for Crown-Indigenous Relations, says he believes the border has had the effect of 'impeding on cultural and traditional activities, which include hunting, cultural exchanges and simple mobility'" (Morin). This alone does not seem enough to exact a change in the laws when it comes to Indigenous people and border crossings. Besides having an impact on cultural activities, "cross-border hunting and the carrying of certain animal parts" is still forbidden, which may have an effect on some Native people's ability to feed themselves, not to mention maintain connections with their cultural heritage. Though the hardship imposed upon Indigenous people by the government in this matter has been acknowledged, nothing has been done to fix the problem by changing the laws. Given the way Canada's First Nations people have been treated by its government over the years, this is unacceptable in today's day and age.

Figure 11a: Two slides of a *Jamboard* multimedia presentation.

THEME

Pride is a good thing to have, you know. Laetitia had a lot of pride, and so did my mother. I figured that someday, I'd have it, too (140).

"Borders" is about having pride in one's personal identity and reveals that people should persist in self-identification rather than let someone bully them into thinking otherwise. On page 139, the customs officer says to the boy's mother: "You have to be American or Canadian." This message has been repeated to the mother each time they try to cross the border but she holds to her self-identity. It makes sense that the boys' mother should be adamant about how she identifies herself, as she would be of the age where she might have suffered through the residential schools, where they tried to strip the children of their identity. Now she is free from the oppression of the schools and living on the reserve where she can practice her heritage. By refusing to pick sides (Canadian or American), she is in effect refusing to allow the government to strip her of her identity.

Figure 11b: Slide of a *Jamboard* multimedia presentation.

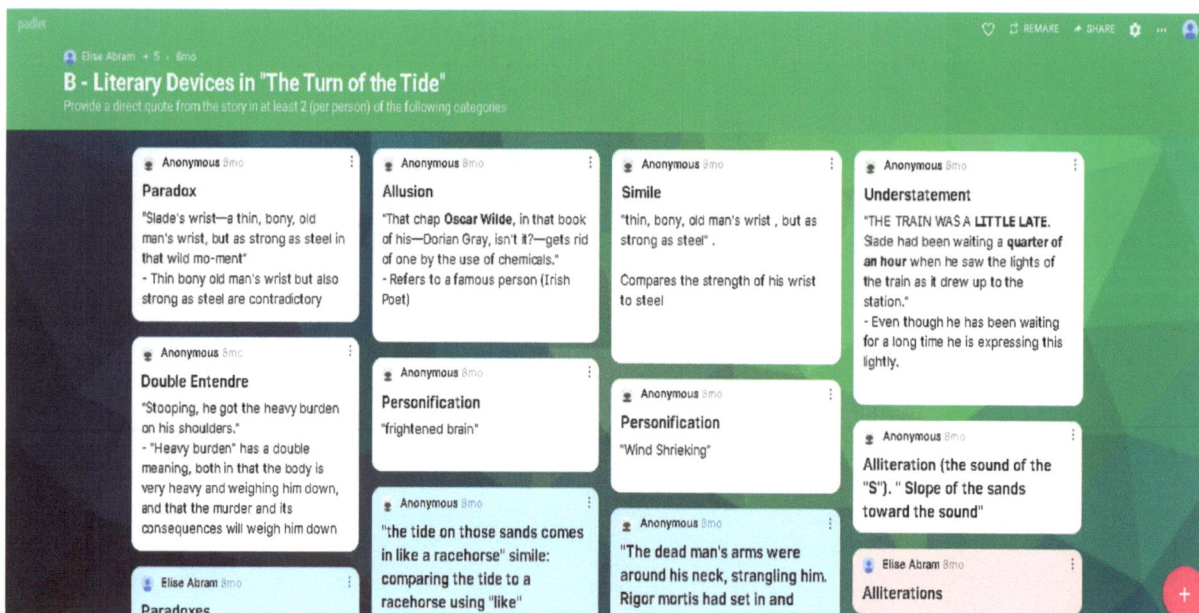

Figure 12: Screen capture of *Padlet* activity.

Some elements students (and teachers) can incorporate in a multimedia assignment using *Padlet*:

- colour (of background or posts)
- images (of background or posts)
- text
- visitor comments
- important quotations and commentary
- adding URLs (instead of embedded links)

Where *Padlet* differs from *Jamboard* is in its organization. With *Jamboard*, students can place posts anywhere on the page, whereas in *Padlet*, information is organized into individual posts spread out on the page as if it were a paper pinned to a bulletin board. One can only post within the area of the "paper" post.

While it is not, technically, a part of the Google Suite, it is integrated with Google, meaning that students and teachers can log in using their Google accounts. Users are allowed a maximum of three free padlets at any given time, which might mean that teachers have to do a bit of juggling if they wish to use the platform in a series of classes simultaneously, although students create padlets using their own accounts and share with teachers, which don't count as one of the three free ones.

Still other app platforms with which students can make multimedia presentations include:

- *Prezi*
- *Canva*
- *Google Slides*
- *PowerPoint*
- *Google Sites*

Alternate forms of multimedia assignments include infographics. According to Midori Nedinger on the *Venngage* website, "an infographic is a collection of imagery...and minimal text that gives an easy-to-understand overview of a topic." As far as infographics go, they have

> the ability to walk you through different phases, offering you facts and intriguing visuals along the way... the flow of [an] infographic [is controlled] using numbers, headers, color [sic], white space, [and] pictures." ("What is?")

They are a way to present more complex information visually and at a glance. In essence, viewers should glean the story of the infographic in a few seconds. If they choose, they may spend more time giving the information a closer look.

While infographics tend to have more visual elements than text, it is kind of different when you are using one for a project in which students must explain their choices, but they work, nonetheless. In keeping with the caveat that you should not be asking students to do anything you haven't previously taught, begin the lesson with a discussion of what infographics are and how they work. I start with a slide deck, giving the same definitions as above and then showing a lot of examples, most of which are taken from the *Venngage* website (although you could do a Google search for "infographics" and pick the ones that most closely resemble those you want the students to create). The examples are interspersed with slides giving advice on how to create an effective infographic, such as:

Infographics are tools for visual communication

"The most visually unique, creative infographics are often the most effective, because they grab our attention and don't let go. But it's crucial to remember that the visuals in an infographic must do more than excite and engage. They must help us understand and remember the content of the infographic." (Nediger)

Infographics make complex information easy to digest

"They can be helpful anytime you want to:

- Provide a quick overview of a topic
- Explain a complex process
- Display research findings or survey data

- Summarize a long blog post or report
- Compare and contrast multiple options
- Raise awareness about an issue or cause." (Nediger)

In short, "[w]hen you need to give someone a really quick rundown on something that can be hard to explain in words alone, an infographic is a good way to go" (Nediger).

I follow this with a slide, also from Nediger's blog post, under the section titled "How do I create an infographic?" Lastly, I share an infographic I created that would be a level 4+ exemplar (see Figures 13 and 13a).

The result is a cross between an infographic and a visual essay.

what's love gotta do with it?

Does love at first sight really exist?
Is Romeo in love or is he in love with the idea of being in love?
Is Romeo just a romantic at heart?

Does love at first sight really exist?

72% **61%**

The percentage of men and women who believe in love at first sight from a 2017 poll (Pomarico).

Studies have shown that people "decide almost immediately if they find someone attractive." True love will not develop without this "initial attraction. Within seconds (or even less), your brain knows if it's interested in who it's looking at, and this can often be what leads to a lasting relationship" (Pomarico). So, it would seem, that Romeo's attraction first to Rosaline and then to Juliet based on their looks is, indeed, the first step to falling in love.

Is Romeo in love or is he in love with the idea of being in love?

How he describes Rosaline

How he describes Juliet

O, she doth teach the torches to burn bright!
It seems she hangs upon the cheek of night
Like a rich jewel in an Ethiope's ear.
Beauty too rich for use, for earth too dear! (I.ii.42-45).

She'll not be hit
With Cupid's arrow; she hath Dian's wit;
And, in strong proof of chastity well arm'd,
From love's weak childish bow she lives unharm'd
(I.i.200-203).

Figure 13: Exemplar infographic created using *Piktochart* template

Mercutio teases Romeo, admitting that he is full of passion, a romantic that speaks in rhyme and is attracted to Rosaline's eyes and body. One has the sense that Romeo has a reputation for being a romantic and using his heart to make his decisions for him, rather than his head. This helps to build the case that Romeo is a hopeless romantic, which is characterized by the way he sees the world. "Hopeless romantics have lofty or elaborate expectations, not just about love, but also about life itself" ("Hopeless"). They are in love with the idea of being in love and being loved back (Roosee). They also often have their head in the clouds (something Romeo is accused of in the play)

The Chorus says...

Now old desire doth in his death-bed lie,
And young affection gapes to be his heir;
That fair for which love groan'd for and would die,
With tender Juliet match'd, is now not fair.
Now Romeo is beloved and loves again,
Alike betwitched by the charm of looks
(II.prologue.1-6).

Explaining that he is finally over Rosaline, but he is only in love with Juliet because of her looks. An article by Dawson McAllister calls a quick connection after a breakup 'infatuation' which may be defined as "the state of being completely carried away by unreasoning passion or love; addictive love...it is characterized by urgency, intensity, sexual desire, and or anxiety, in which there is an extreme absorption in another." This sounds a lot like the relationship Romeo and Juliet have.

Is Romeo just a romantic at heart?

Isn't it romantic?

Romeo! humours! madman! passion! lover!
Appear thou in the likeness of a sigh:
Speak but one rhyme, and I am satisfied;
Cry but 'Ay me!' pronounce but 'love' and 'dove;'...
I conjure thee by Rosaline's bright eyes,
By her high forehead and her scarlet lip,
By her fine foot, straight leg and quivering thigh
And the demesnes that there adjacent lie,
That in thy likeness thou appear to us! (II.i.6-20).

Mercutio teases Romeo, admitting that he is full of passion, a romantic that speaks in rhyme and is attracted to Rosaline's eyes and body. One has the sense that Romeo has a reputation for being a romantic and using his heart to make his decisions for him, rather than his head. This helps to build the case that Romeo is a hopeless romantic, which is characterized by the way he sees the world. "Hopeless romantics have lofty or elaborate expectations, not just about love, but also about life itself" ("Hopeless"). They are in love with the idea of being in love and being loved back (Roosee). They also often have their head in the clouds (something Romeo is accused of in the play)

Friar Laurence Knows

Holy Saint Francis, what a change is here!
Is Rosaline, whom thou didst love so dear,
So soon forsaken? young men's love then lies
Not truly in their hearts, but in their eyes.
Jesu Maria, what a deal of brine
Hath wash'd thy sallow cheeks for Rosaline!
Thou and these woes were all for Rosaline:
And art thou changed? pronounce this sentence then,
Women may fall, when there's no strength in men (III.ii.69-85).

Figure 13a: Exemplar infographic created using *Piktochart* template

5. COLLABORATIVE NOTE-MAKING

Because Web 2.0 apps may be accessed anywhere and anyone with permission may edit documents, they are excellent tools for note-making, be it teacher-led or collaborative between students. See above for some examples of teacher-led collaborative notes.

Jamboard is a great tool for facilitating collaborative student notes. It's a great substitution for asking select students to write their answers on the board—if you've ever done this, you know that trying to get students to write their answers on the board can be like pulling teeth. When using Web 2.0 tools, it is easier to encourage student engagement as the results are semi-anonymous, and students are not drawing attention to themselves—and potentially, their mistakes—by making them in front of the class.

Collaborative notes in *Google Docs* can be great for completing student tasks in small groups. To set this up, create a single document for each group and pre-load the questions or worksheets to be answered. As each student has access to the file, they may divide up the work as they see fit. Students can see their peers' answers in real-time, and they may decide to use the comments feature to communicate with each other.Students may also decide to each use a different colour so they can track who contributes what.

Collaborative notes may also be made using *Google Slides*. For example, the teacher can begin the slide deck by pre-loading slide 1 with a title page, slide 2 with instructions, and slide 3 with an exemplar. After the side deck is introduced to students, time is given for each student to complete his/her own slide, creating a set of collaborative notes with a sample of all of the students' work in one place.

6. WEBQUESTS

Webquests have been around for almost as long as teachers have been using the Internet in education. "'A WebQuest,' according to Bernie Dodge, the originator of the webquest concept, is an inquiry-oriented activity in which most or all of the information used by learners is drawn from the Web. WebQuests are designed to use learners' time well, to focus on using information rather than on looking for it, and to support learners' thinking at the levels of analysis, synthesis, and evaluation. (Starr)

In short, webquests are web sites created to lead students through an inquiry-based task, such as the process of investigating the context of an author of a work of fiction. When creating a webquest, the teacher

- plans out the lesson;
- breaks it into smaller tasks;
- finds links online that will help guide the student and/or provide information on which the teacher wants to focus;
- gives students instruction as to what to do with the information
- once they have reviewed it, such as answering an inquiry-based question; and
- makes the website look inviting, entertaining, and interesting by adding images, colour, and changing font size.

Webquests may be created using a pre-existing template, although starting from scratch using *Google Sites* might be better, as many templates won't necessarily suit the teacher's needs. Once a teacher has created and used a webquest with her students, she can demonstrate how it was made using the software and then suggest it as a platform for students to use when creating their own multimedia assignments.

7. QUIZLET

Another great way to create diagnostic and formative quizzes is using *Quizlet*, a flashcard app that allows you to sample other people's flashcards as well as create a "stack" of your own. This is a quick way to poll students to determine their understanding. I usually show the flashcard and ask students to show their answers by raising their hands before revealing the answer.

In the flashcard deck I used at the end of my lesson on critical inquiry questions, I showed students a question (for example, "How many Canadian cities have COVID cases?"). As questions are either open-ended, closed-ended, or opinion-based, after sharing this question, I asked for a show of hands to vote for the type of question they thought it was, before clicking on the flashcard to reveal the answer.

As an added bonus, you can send students to *Quizlet* to make a series of questions with which they can test themselves and other students. We often ask students to do seminars instead of simple information-disseminating presentations, and *Quizlet* is a good tool for them to incorporate into their seminars to introduce an element of interactivity and test which students have been paying attention. Card decks are available to your students all the time, so you can also create a set of review questions and have students complete them at their leisure.

Quizlet is free, and if your school has a subscription to *Google Apps for Education*, you and your students can use your Gapps ID to login and get started. The only downside is that you must have a paid account for completion statistics, but if I need to keep a record for student accountability, I will use *Google Forms* instead. Still, as mentioned previously, sometimes it is good to shake things up in the classroom to keep them fresh, and *Quizlet* is an easy-to-use alternative to Gapps.

GAMIFICATION

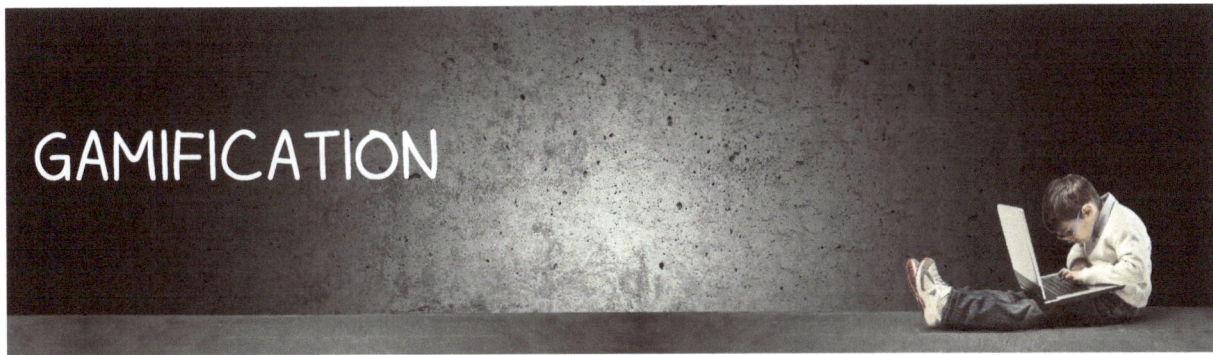

8. ESCAPE ROOMS

Traditional escape rooms see people physically locked in rooms with the task of "uncover[ing] hidden clues, crack[ing] tough codes, and solv[ing] challenging puzzles" to find a way out. Similarly, digital escape rooms present students with a series of puzzles, each of which leads them to a "key" and another puzzle until they "escape" the "room," having solved each of the puzzles successfully.

I thought that creating a digital escape room would be a great way to consolidate my elements of storytelling analysis unit, but I had no idea where to begin. Then I found Meredith Dobbs's most excellent instructions in a blog post on the *Teach Writing* website, entitled "How to Build a Digital Escape Room Using Google Forms." I even purchased a set of puzzles from her on the *Teachers Pay Teachers* website (link available in the above mentioned post) and modelled my puzzle after hers. The instructions are step-by-step and easy to follow.

The result was that my students were engaged for anywhere from 30 to 45 minutes, and many would have taken more time had I let them, based on their skill set. My series of puzzles was five long and focused on identification and matching (match the literary device to its definition or to an example), scavenger-hunts (find author and contextual information online), theme labelling, and the like. Each of the puzzles yielded a series of letters forming the key to the puzzle and taking the students to the next level or "room." I offered the winning team—the first one to successfully "escape"—a prize (a pencil, eraser, pen, marker, notepad, etc., I had purchased from the dollar store in a bulk package so each prize cost between ten cents to a quarter to purchase) as an incentive.

Some students were frustrated when they couldn't get the correct sequence of letters based on their answers. At that point, I stepped in and listened to their answers to point out where they had gone wrong so they could refocus and reorganize, and they were, eventually successful a short time thereafter.

All in all, the morning it took to create the escape room and tailor the puzzles to support my curriculum was well worth the time to see how engaged the students were in the results.

9. QUIZIZZ AND KAHOOT!

Both *Quizizz* and *Kahoot!* are similar tools in that they allow students to log in to participate in real-time quizzes that are scored based on the speed of the answer and selecting the correct one. The apps keep track of the top-scoring students and display them and the answers to the questions once the time set for answering the question runs out. Both platforms allow teachers to assign their quizzes as homework for students to complete on their own time or to be used in a teacher-led lesson during class.

Both sites:
- allow teachers to revisit quizzes, individual student results, and statistics
- engage students by gamifying diagnostic and formative assessment
- allow students to work through quizzes independently or as a teacher-led lesson
- have free and paid versions

Quizizz seems to have more free features, including:
- allowing teachers to create slides without quiz elements
- allowing teachers to upload existing slide decks to integrate into the lesson
- displays the entire question as well as selection buttons on the student's monitor
- gives teachers multiple question options besides simple multiple-choice, including fill in the blank (short answer), and ungraded poll and longer answer questions.

In my time, seeing the teacher wheel in the television and VCR elicited the same oohs and ahs the *Kahoot!* and *Quizizz* setup screens do today, and thanks to the interactivity and gamification that comes with these two sites, you can be sure the lesson will be much more engaging.

TED-ED

The *Ted-Ed* web site is the epitome of Web 2.0 when it comes to "user-generated content," as it allows you to use any of the already created lessons for free in addition to customizing and re-sharing any of the lessons you find. Each of the existing *TED-Ed* lessons is divided into four parts:

- *Watch*: a "minds-on" video
- *Think*: a series of multiple-choice and/or open answer questions
- *Dig Deeper*: providing students with additional information and/or "resources to explore"
- *Discuss*: a forum for students to discuss an open-ended or guided question with each other.

There is also a "Customize" button, easily allowing teachers to modify any and all blocks of the lesson, including the minds-on video. When done, teachers have the option to turn the customization for future users off, if they choose. As previous users have already culled potential video choices for you, it means you don't have to spend the time, but you can modify the lesson to correct for variable terminology (changing "topic sentence" to "claim," for example) and/or tailor the questions to suit the expectations you are hoping to impart with the lesson.

You can also create a lesson from scratch, track who has completed the task and see student answers from the teacher dashboard. While none of this is new—you can do the exact same thing using *Google Apps*—sometimes it is better to shake things up a bit and send students to an interface other than *Google*, so they perceive it as something fresh or different. For more on how to set up a *TED-Ed* lesson and the kinds of information a teacher receives once it is done, watch Russell Stannard's video, "How to use TED-ED—Ideas for the Flipped Classroom."

10. BOGGLE AND OTHER GAMES

In the English language classroom, playing games like *Boggle* does double-time as it works on students' vocabulary and spelling skills. To play *Boggle* as a class, select your choice of puzzle. The *Puzzle Words* website offers a number of formats. Once you have your puzzle, show it to the students by sharing your screen for online students and/or with an LCD projector for face-to-face students. Decide how much time you will spend on each puzzle (I chose five minutes), and tell students to write any and all words they see on a separate sheet of paper or a new document. Remind students that while words may be vertical, horizontal, diagonal, or a combination, letters must appear adjacent to each other in the order they appear in the word. Once you have reached your pre-determined time, go around the class, asking each student to contribute one word from their list, which you can type in the list space to the right of the *Boggle* board. Students should cross out the word if they have it on their lists. The winner is the person who has the most unique words on his/her list. Again, you may choose to offer a Dollar Store prize to the winner or the top three winners. Games go fast, so depending on how much time you have set aside for the task, you might be able to complete two or three rounds.

Other games you can play as a class with your students include:
- *Jeopardy* review (templates abound online for both *PowerPoint* and *Google Slides*;
- *Family Feud* (likewise, templates abound);
- *Hangman* (find an online version and students can guess letters one at a time or guess the whole word);
- rhebus puzzles (can find many different ones online); and/or Do memory games in which you display a collage of images and words for two to five minutes, after which, students are given two to five minutes to write down all they remember seeing. The winner is the student with the most items on his/her list. A few great ones (including some rhebus and other puzzles)can be found in Kathy Schrock's *The Critical Thinking Workbook*.

11. POINT AND CLICK GAMES

With a little coding know-how, it is easy to make simple concept attainment tasks from scratch to gamify your lessons. Here is a task I created while teaching Information and Computer Studies using JavaScript (see Figure 14).

Rather than lecture students on the parts of a URL, I created this point-and-click "game" to do it for them. A description of the part of the URL appears in the window below the colourful web address in which each colour represents a different component of the uniform resource locator (URL). Each of the parts is a clickable graphic. Students click on the graphic they think best exemplifies the description given, and the website records how many turns or questions the students have addressed, how many correct guesses they've made, and it gives them an accuracy percentage. Not only does this gamify the lesson, but it adds a competitive dimension as students can compare their accuracy percentage with their classmates. If students want to improve their accuracy scores, they just have to reload the page to reset the form.

As for the technical aspects (if you know JavaScript), the game uses a single form and a set of linked arrays, one storing the correct answers and one storing the descriptions. Each graphic returns a number when clicked, and if the statistics are adjusted accordingly, depending on if the number returned by the form matches the answer in the array. Though the form doesn't currently collect student results, this can be easily modified by having the student put his/her name into a separate text field in the form and emailing the teacher the contents of the name field along with the bottom three fields displaying the statistics when they refresh the page.

Parts of URL Matching Game

Click on the part of the URL that matches the description in the text field below.

http :// www / askteacheronline . com / ICS2O / Asst1 . html

A file on the server.

Turns: 1   Guesses: 0   Accuracy:

Figure 14: Screen capture of teacher-coded matching game

As an extension to the lesson, the teacher can ask the students to create a similar game on their own. I begin by sharing the code for this program, breaking it down step-by-step, and having them recreate the code on their own according to my instructions (for computer programming classes and depending on their grade and JavaScript know-how). Possibilities for alternate, similar tasks include filling hamburger orders (clicking on the graphics in the order in which the customer places the order), filling pillboxes (three green pills on Monday, one green and one red on Tuesday, etc.), and mathematical car races in which a car moves forward along a track for every math question the user gets correct and their competition moves forward for every question the user misses.

This is a perfect Web 2.0 tool as not only does it allow the user to interact with the software, it gives students the opportunity to master the concept and provides an extension to teach students how to contribute to the Internet by creating their own apps.

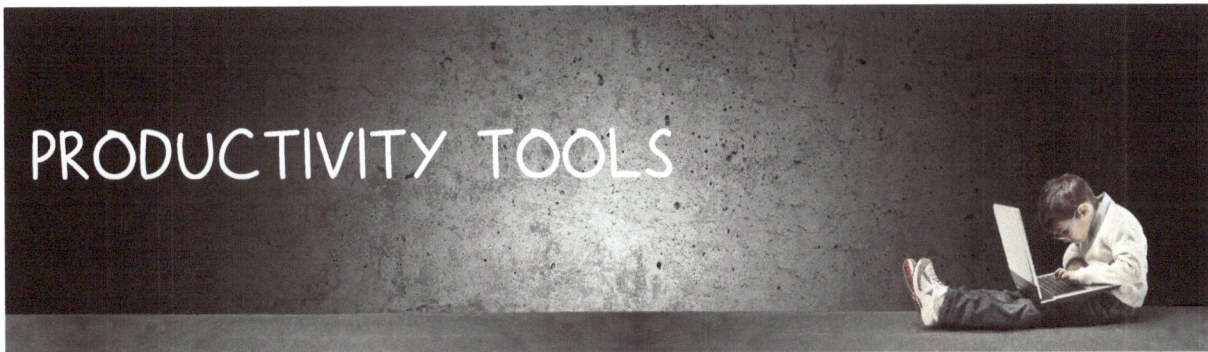

PRODUCTIVITY TOOLS

12. GOOGLE AS A LEARNING MANAGEMENT (LMS) TOOL

One of the great aspects of Web 2.0 is that it provides access to a number of excellent productivity tools to help keep teachers and students organized and efficient.

In his article "The Possibilities of Web 2.0 and Google Apps," Joe Zappa calls *Google Apps* "a version of a learning management system" (LMS) that is not as "daunting" as other, more complex LMS. In addition to being collaborative and interactive, *Google Apps* allows teachers and students to stay organized and exchange assignments and feedback (Zappa). Because Gapps may be accessed anywhere in the world and on any device (Zappa), it facilitates efficiency on the part of both student and teacher.

Google applications are excellent when it comes to student tracking and accountability as you can trace virtually everything a student does while online in the interface. Here are a few ways to use *Google* Apps that you might not have thought of before.

I used *Moodle* as my LMS for the better part of a decade, but once we switched to emergency online learning, I thought it best to give *Google Classroom* another go, as it is a one-stop-shop for most online classroom management tools. To *Google's* credit, they worked with teachers at the beginning of the pandemic to ease the growing pains teachers, students, and the *Google* LMS experienced to make their products even better than before. Here is an overview of some of the many *Google* apps that form the LMS as well as suggestions for how to incorporate them into lessons for curriculum delivery and enhancement.

Email

For years, the Ontario Secondary School Teachers Federation (the teacher's union, the OSSTF) and the Ontario College of Teachers (OCT, Ontario's licensing body for teaching certificates) have been urging teachers not to email students. To switch almost exclusively to communicating with students via email was a huge about-face, but *Gmail* is a good compromise. When a board has a license to use *Google Suite*, it includes *Gmail* accounts, and the entire system is closed to all but board staff and students. If discussions with students must take place via email, shunting all communication through *Gmail* (rather than a board account) seems like a good idea.

One of the good things about *Gmail* is that it streams the address process, suggesting addresses of people attached to your board's license. It is also intuitive in that it learns which students are in your class, and it suggests a list when you begin typing a student's (or teacher's) name. While the first time you might have to search from a list of all potential students with the same name, once you select an email from the list, *Gmail* "learns" that this is the person you mean, and it suggests that email first the next time you start typing that particular combination of characters, making sending multiple emails a breeze and negating the need to remember emails composed of student numbers.

Even if a board doesn't have a *Google* license, *Gmail* accounts are free as are other *Google* products, so teachers can still use them with their students (although if there is no license for use, it is recommended teachers seek parent approval before using *Google Suite* of tools). It should be noted that this is not meant as an endorsement for *Google* apps. Rather, it is meant to demonstrate my best practices when incorporating Web 2.0 apps in my teaching practice, and since my board does have a license, many of the tools I use are in the *Google* suite of tools.

Classroom

Google Classroom is a part of the *Google Suite* applications, and it has come a long way since it was first introduced. To *Google*'s credit, they are open and responsive to suggestions, and they have made a number of modifications over the past year to accommodate safe and secure classroom management.

Google Classroom is primarily a message board on which teachers and students can communicate, like a cross between *Facebook* and *Twitter*. Teachers can post announcements to the classroom stream, and students can respond or ask questions via the interface (although the ability to respond can be turned off if teachers wish). There are a number of other excellent features embedded within the *Google Classroom* app to help facilitate teachers and students in the day-to-day machinations of the classroom. Some of them are:

- *To-do lists*

Whenever an assignment is entered, it appears on a personalized to-do list for each student, allowing them to keep track of their progress.

Note that at the time of publication, teachers cannot see the student interface in *Google Classroom*, but the students' to-do list is similar to the teacher's "To review" list.

- *Calendar*

Assignments posted with due dates automatically show up on a calendar, letting students know of upcoming assignments. The nice thing about this is that if multiple teachers use *Google Classroom*, assignments from all of a student's classes will show up on the calendar (and the to-do list, too), which is great for students to plan their study time.

- *Materials*

Teachers can post BLMs to the materials section. They can organize the materials according to bespoke categories. Notes as to further instructions can also be left for students. The materials section is easy to organize and reorganize as the semester progresses. Notices that the teacher has posted a resource automatically appear in the classroom stream to notify students.

- *Quizzes*

Teachers can create and administer tests right in the *Google Classroom* app. This function is the same as the quiz function in *Google Forms*, and it has the same features. Teachers can ask multiple-choice, true/false, fill-in-the-blank, and short and long answer questions (and a few other types), and some of them (like multiple choice or true/false) are automatically graded. Questions with answers that will vary (like short and long answer questions) still require manual grading.

- *Accepting work*

Google Classroom allows students to submit their work. This is keyed into the *Google Docs* app, meaning that once an assignment is submitted, teachers have access to all *Google Docs* features for annotating and commenting on student work. Teachers can:
  - see the date of submission;
  - leave a private message for the student;
  - respond to students' private comments;
  - see who has submitted and who hasn't at a glance;
  - open documents in *Google Docs* to edit and/or comment; and

- detect plagiarism (of Internet sources—unlike *TurnItIn* and similar software, *Google*'s plagiarism detection software does not maintain a database).

There is also an integrated gradebook for boards that do not already use similar software.

Once work has been uploaded by students, the documents are frozen until the teacher can evaluate it, which is good for taking a snapshot of student progress, although students can make a copy of the file and continue working on it. When teachers are done evaluating the student's work, they simply click "Return" to send the work back to the students and "unlock" the file.

- *Forms*

Google Forms has a number of templates included in the app, but it is really easy to create and customize your own files. The interface works similar to other *Google* apps, so it is really easy to navigate. The nice thing about *Google Forms* is the way it presents information to teachers, including:
- Individual student responses.
- A summary of all responses. This is particularly useful with multiple choice or true/false questions, as the interface creates a bar graph to show amalgamated student results.
- A spreadsheet of responses. Generating and saving a spreadsheet containing all student responses is a great way to store data collected for future reference.

Some ways to use *Google Forms* in the classroom include:
- *Exit tickets* (See section on exit tickets above)
- *Polls*
  For wellness check-ins, to see how comfortable students feel with the current lesson, and/or opinion polls, among others. These are great because data can be collected in real-time and bar graphs generated by the interface can be shared with the class as a whole, which can spark further conversation.
- *Information collection*
  For example, at the start of a semester to collect contact information and gauge an idea of student achievement in previous years. This can also be done during a lesson to have students create lists of resources or brainstorm.
- *Resource registration*
  To declare essay topics and/or culminating activity texts or to sign out books.
- *Quizzes* (graded and ungraded).

Jamboard

As previously mentioned, *Jamboard* is a digital whiteboard app available in *Google Suite* and integrated in *Google Meet*. Teachers can create as many jamboards as they would like for free, and they may be saved for posterity, copied to reuse, or student posts can be cleared for the actibity to be reused at a later date. One great feature is that teachers can upload images as backgrounds for their whiteboards, and students can write and/or post on top of the backgrounds, allowing teachers and students to annotate and/or comment on the document.

There are a number of ways you can use *Jamboard*, such as:

- *Polls*
  Asking students to post a sticky-note-like icon to vote for their answer as if in a wellness check or to gauge comfort with a lesson. (See also Critical Inquiry section).
- *Dot-ocracies*
  In a dot-ocracy, students are given a number of choices and they "vote" on their choice(s) using a sticky-note. Students may vote once or any number of times based on the purpose of the lesson.
- *Taking up work*
  By uploading an image of the worksheet (for example, in the Freytag pyramid or organizer; see section on collaborative note-taking and -making), teachers can ask students to complete the worksheet collaboratively with other students in the class.

13. WORKS CITED LISTS ON THE FLY

One of the hardest things for students to do, for some reason, is to generate properly formatted works cited lists. *EasyBib* provides a useful free solution for this (although the number of ads on the site is maddening and distracting). When using *EasyBib*, students type in the title of their source (for print documents) or URL (for web documents), and the interface returns a list of potential sources. Students choose the source they used, fill in missing information, if available, and the interface adds it to a works cited list students can "export" or copy-and-paste to *Word* or *Google Docs* (Note that Word and GoogleDocs also have the ability to create works cited listson their own). *EasyBib* also has a great store of information, detailing everything you will ever need to know about a variety of citation formats that might prove useful for student use.

14. GRAMMARLY

Grammarly allows you to upload or copy and paste your documents into its interface so it can suggest what might be errors in grammar, sentence structure, punctuation, word choice, and tone. It also has a plagiarism checker that compares your document to others it finds on the Web. The plagiarism detection isn't often accurate, as most common phrases used link to websites that have nothing to do with the content of your manuscript, but when it does find something, it provides you with a citation to the original document.

Grammarly is a nifty tool to help polish your writing. It has both free and paid versions, with the paid version offering more suggestions, but speaking as someone who uses the Premium version regularly, too many suggestions can be overwhelming. Also, if you don't know English grammar rules, you may wind up accepting some suggestions that are not appropriate. If you remember that Grammarly is an AI and not a person who can weigh if a change makes sense, you should be okay using it.

15. DIGITAL TEACHER'S PLANNER

There are a number of sites out there offering both free and paid digital teacher's planners, but some of them require the installation and/or use of a separate piece of software (such as *Microsoft One Note*). Others have fixed background images, which look nice until you have to enter an agenda or note longer than the space provided by the background. The best thing is to create your own digital teacher's planner using *Google Suite*. That way, you can tailor it to your needs.

To begin creating a digital teacher's workbook, create a new folder in *Google Drive* and call it something like "teacher planner." Any time you create or add a record, save it in this folder. Create a number of sub-folders to store similar information for multiple classes. Some categories that might help with record-keeping include:

- *Washroom logs*
The original template for this was provided by my board as a *Google* doc. I keep a separate one for each class, and we have to upload the log at the end of each face-to-face block for

contact-tracing purposes.

- *Assignment tracker*

This takes the form of one spreadsheet per class. On it, track each time a student participates in a collaborative activity (in *Jamboard*, for example), as well as which assignments have been received and when. Append comments to cells to keep anecdotal notes about individual students' assignments, tasks, or performances. Create a second spreadsheet in the same document to chart accommodations necessary for individual students with IEPs (Individual Education Plans).

- *Notes*

Keep a *Google* doc handy for recording anecdotal notes for conversations with students or contacts home. Alternately, you could keep a separate document or spreadsheet to record parent-teacher contacts via phone or email.

- *Long-range calendar*

Create a table in a *Google* doc on which to record your long-range calendar. Include
- which classes are digital or face-to-face,
- the length of each class,
- scheduled breaks,
- daily learning targets, agendas, and success criteria, and
- link agenda items to resources needed for the day's lesson for easy access.

16. FAQ PAGE

Due to its collaborative nature, an FAQ page is a great reference tool to create for students. It also cuts down on the number of times you have to answer certain questions ("Are you going to post this in the *Google Classroom*?"). I found that this works well, provided students contribute. I was also selective in what I chose to answer, as questions of a more personal nature that were specific and probably better to answer one-on-one received an email in response.

How it works

Create a new Google Doc page with the following message on it:

How to use this page:

Check the list below to make sure your question hasn't already been asked and answered.

If your question isn't already on the list, type your question below.
I will check this page once a day to answer any new questions that might have been posted.

Check back on this page to find your answers.

Note: Please DO NOT answer questions from your peers here. This page is strictly for me to answer your questions.

I created a list with the header "New Questions (posted by students)" and posted the first question myself (Will the materials you use in class be posted to the *Google Classroom*?). Below that, I used the header "Questions and Answers (posted by students)" under which I re-wrote the question and gave my answer. For easy access, I linked the question at the top of the page to the answer at the bottom using a bookmark. Every time I made a change to the file, I posted a note in *Google Classroom* feed that the file had been updated, and I listed the questions I'd answered.

To grow your list of FAQs, consider asking students for questions they have about the last lesson or unit on an exit ticket and cull questions from there to add to the page.

17. TURNITIN AND PEERMARK

For close to the last fifteen years, the English department at my school has insisted that assignments be uploaded to *TurnItIn* for plagiarism detection prior to being evaluated. For those who have not yet used this Web 2.0 tool, *TurnItIn* takes students' work and compares it against everything on the Internet, anything that has been published digitally, and anything that has ever been uploaded to its database and returns a "similarity index" indicating the uniqueness of the paper. It is then up to the teacher to review the paper to ensure that what has been highlighted by the interface is, indeed, plagiarized (i.e., copied and pasted verbatim from another source and not cited). Students are able to see their similarity indices, allowing them to revise and resubmit if they choose. In the early days, I asked students to submit their work twice, once to *TurnItIn*, exclusively for plagiarism detection and once as a hard copy for feedback, but for the past five years or so, I have required them to upload exclusively to *TurnItIn*.

The teacher dashboard allows you to open, read, and review files. You can append files by typing directly on the page or by adding comment bubbles. There is even a bank of comments you can use and add to, personalizing your editing codes as you go. When you are done, students can log back into *TurnItIn* to view your comments.

The nice thing about this is that student submissions and your comments are left in your account for its duration, so I could go back to review my comments on a student's paper submitted years ago if I choose. Also, the interface connects all teachers using *TurnItIn* to get to the bottom of plagiarism issues. If a student assignment turns up as plagiarized from another student—be it in my school or anywhere in the world, for that matter—*TurnItIn* gives me the option to contact the teacher, asking for permission to see the original document in entirety. All the other teacher needs to do is click a button, and the original document is emailed to me for my perusal. If a document turns up plagiarized from a former student of mine (as it often does), it automatically releases the original document in its entirely for my comparison.

As a part of your *TurnItIn* subscription, you have access to *PeerMark*, an app that helps facilitate peer editing. Depending on how you set this up, you can assign students a number of documents to edit, including their own. The last time I did this, I required students to view two of their peers' assignments as well as their own. Once uploaded, the interface distributes the assignments at random. All students need do is log into *TurnItIn*, click on the assignment title, and begin editing.

Setting up the *PeerMark* activity takes some time. After reading and editing their peers' work, which entailed typing notes directly on the paper, similar to how I might when evaluating their work, students were asked to answer a series of questions for the author of the paper. These questions had to be entered in the interface prior to the tasks assignment and included:

1. The thesis is a three-part thesis.
2. The claims function as sub-topics of the thesis.
3. The essay is in MLA format.
4. The essay is in formal tone.
5. Introductory paragraph follows structure given.
6. All paragraphs are CLEER format.

7. Signal phrases are used to transition to quotes.
8. MLA in-text citations are used throughout.
9. Correctly formatted works cited list.
10. Structure of conclusion.
11. Spelling, grammar, punctuation.
12. Identify two strengths of this draft.
13. Make two specific suggestions for improvement.

Questions one through 11 are graded on a scale of one to five, with the last two questions requiring short paragraph answers. These questions could be altered to address whatever aspect of the assignment you would like to highlight. When done, students have access to at least two other students' opinions as to the efficacy of their assignments.

This worked really well for the students who uploaded their essays prior to the beginning of the task. Those who chose to upload their assignments after the assigned due date had to wait for their essays to trickle through the system, and some of them were unable to participate in the task as a result. This is undaunting as even when the peer editing process is carried out using paper and pencil, there are still students who are unready to participate in the task and must look elsewhere to complete the peer editing portion of the process. It is, however, important to note that this is one of those tasks in which the teacher must ensure there is sufficient technology to go around.

18. DIGITAL RUBRICS

One of the biggest inconveniences when collecting student assignments is ensuring the rubric comes back to you when it is time for evaluations. Short of printing out twice as many copies of the rubric as you have students (which was my practice for a very long time), I have switched to digital rubrics.

BLM rubrics work well if the student includes a copy and paste of the original rubric distributed in the *Google Classroom*, but when they don't, my go-to has been to open the file myself and copy and paste the band descriptor as needed, but that is time-consuming, as the formatting usually doesn't port over to the *TurnItIn* interface, and you cannot add pages to the file uploaded, so your space for commenting is limited. As a solution, I decided to switch to digital rubrics.

It should be noted that both *TurnItIn* and *Google Classroom* have the ability to create digital rubrics in their interface, but I have not yet been able to figure either of them out. Then I read Matt Miller's blog post, "25 practical ways to use *Google Forms* in class, school" on his *Ditch That Textbook* website. Under the section "Assessment: 15. Rubrics," Miller suggests creating rubrics in *Google Forms* and using "the *Autocrat* add-on to turn all that feedback into a document [and s]hare that document with students (or parents too!)."

My rubric has four sections, one for each of Knowledge, Thinking, Communication, and Application, per the grades nine to 12 achievement chart. We score on a scale of one to four, with the option of plusses and minuses. Not meeting the expectations is scored as remedial (R). I chose radio boxes for my forms instead of multiple choice, as this will allow me to tick multiple boxes (for example, to indicate a grade of 4+). Each category also has a comments section.

When setting up the form, I chose to automatically collect the user's email address and return a copy of the answers to the user. In this case, it will collect my address each time and send a copy of the answers to my *Google* account, which I can then forward to the student in question. Though it was time-consuming to copy and paste each descriptor to set up the *Google* form, I hope this will expedite the process when it comes to assignment evaluation over the long term. As an added bonus, I can create a spreadsheet of all the rubrics for later reference, so there is no relying on students to keep rubrics if I forget to record a mark, record incorrectly, or the student has questions or challenges about the way the assignment was marked but has neglected to keep his/her rubric.

19. METACOGNITIVE EVALUATIONS

A good portion of the English expectations in the Ontario curriculum are based on metacognitive assessment and evaluation. A while ago, our department realized that though this was a sizeable percentage of our course expectations, we were not all formally evaluating—or even requiring—students to reflect on their learning and skills. The solution was to formalize a full five percent of our marks under the thinking category devoted to metacognitive assessment. We created a rubric and set out to incorporate the evaluations.

One way of integrating this into the curriculum is to get students to write personal responses. Another way is to complete "goal envelopes" in which students complete an organizer celebrating past performance and penning action plans for improvement. The results are folded up and sealed like an envelope and given to the teacher for safekeeping. After the next assignment, students are given back their goal envelopes, open them, and respond on how or if they achieved their goals, and the process begins again.

It soon became apparent that metacognitive assessments are strictly for the student's reference and use, and the teacher need not evaluate the process except for completion. That was mind-blower number one. Mind-blower number two came with the normalization of Web 2.0 tools like *Google Forms*. The best way to assign metacognitives is to put the questions in a *Google* form and ask students to complete them there. This way, the teacher can easily track their completion as well as know if students have completed the task at a glance. Longer questions, such as action plans, might warrant a teacher's skim of what the student has written to ensure they are able to support students with their needs moving forward.

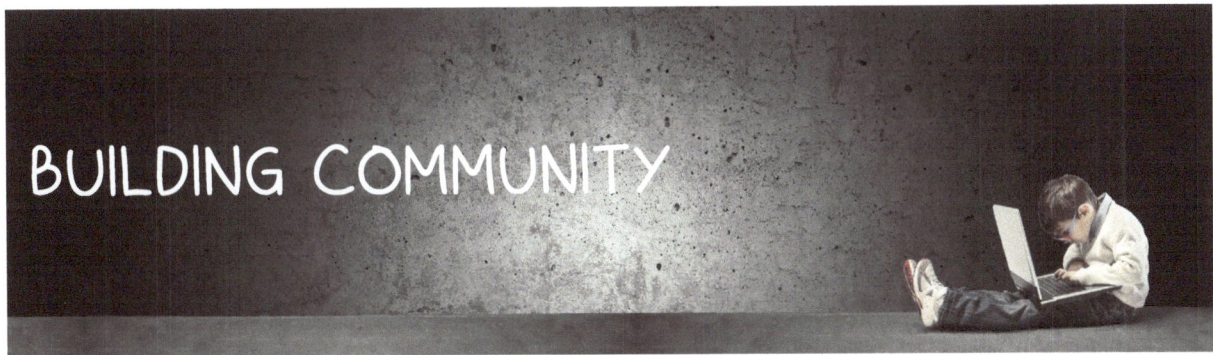

BUILDING COMMUNITY

Web 2.0 tools are great for icebreakers and activities to build community in the classroom. These tasks, done online, crunch student responses to give real-time results with which students may compare themselves with and learn something about their classmates. Some examples include:

20. POLLS

Using *Google Forms*, you can share any student answers with the rest of the class. Longer answers are listed in the order in which they were posted and without student names. The resulting anonymity takes away any chance of students feeling embarrassed if their answers don't add up to their peers'.

21. INTERACTIVE COLLABORATIVE SLIDE DECKS

Pear Deck is a *Google Slides* add-on with free and paid options. In the free version, teachers have access to four types of questions (text answers, choice, number, or website). *Pear Deck* works similar to *Kahoot!* and *Quizizz* in that when you run the program, it generates a passcode for students to use. Once students go to the *Pear Deck* website and enter the passcode, your slides are broadcast onto the students' screens. Students can contribute to the slide polls on their own devices, and the answers may be accessed on the teacher's computer. The only drawback is that the free version only allows you to see student answers while the slide show is active. When a slide is active, students may draw anywhere on the slide using their mouse or finger if they have a touch screen. The answer is collaborative, so all students' answers show up on the same slide. Other slides are individual to the student, but all slides may be viewed simultaneously by the teacher (and shared with other students) while the slide show is active.

22. POLL EVERYWHERE

Poll Everywhere allows you to create dynamic words clouds with which to poll your students. The resulting word cloud may be shared in class in realtime. On *The Resilient Educator* website, Monica Fuglei defines word clouds as: "a picture made from a piece of text...generally created from an algorithm that...measures word frequency and represents word size based on the number of appearances in the text." While there are many sites out there that create word clouds, I chose *Poll Everywhere* for its ease of use. The question I posed my students as a getting-to-know-you task was "When you were a child, what did you want to be when you grew up?" The system generates a URL with which students can access the interface to contribute to the word cloud, which is updated in real-time.

ABOUT THE AUTHOR

Elise Abram has been a high school teacher of English and Computer Studies for close to 25 years, providing coaching to writers of all ages and at all levels of development from middle school through Ph.D. candidates. She is a former archaeologist, and current editor, freelance writer, award-winning author, avid reader of literary and science fiction, and student of the human condition.

Abram is best known as a contemporary young adult novelist, but her writing interests are diverse. She has published everything from science fiction for adults and young adults to paranormal fiction for adults and young adult paranormal, police procedural, and young adult contemporary. She has also published four children's picture books.

In 2015, Abram formalized her company, EMSA Publishing, in order to edit and publicize the work of other authors. She also freelances as a writer, editor, cover designer, and book formatter.

Prior to becoming a writer, Abram worked as an archaeologist in the Greater Toronto Area and across Ontario for ten years, excavating a combination of prehistoric, contact, and historic sites. She has organized and designed curriculum for an archaeological field school and camp groups at the Royal Ontario Museum, where she was a teacher for three years.

Abram holds a B.A. in Cultural Anthropology from the University of Waterloo in Ontario, Canada; a B.Ed. in Learning in Non-School Environments with a focus on Archaeological Education from the Ontario Institute for Studies in Education, the University of Toronto, in Ontario Canada; and additional educational qualifications in subjects ranging from Contemporary Studies to Library Studies.

LIST OF WORKS CITED

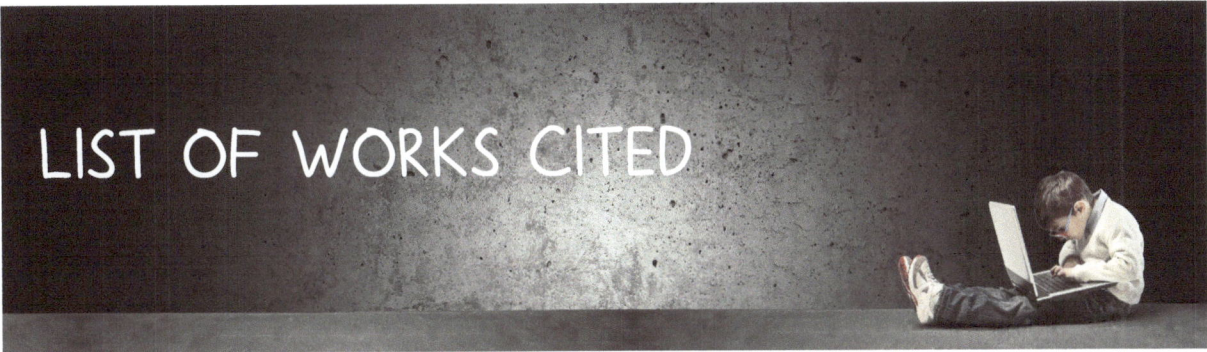

"24 Printable Exit Slip Templates." TemplateLAB, 2021. https://templatelab.com/exit-tickets/. Accessed 30 Oct. 2021.

An, Yun-Jo, Bosede Aworuwa Glenda Ballard, and Kevin Williams. "Teaching with Web 2.0 Technologies: Benefits, Barriers and Best Practices." AECT: Association for Educational Communications & Technology, 2009. https://members.aect.org/pdf/Proceedings/proceedings09/2009/09_1.pdf. Accessed 28 Oct. 2021.

Boyd, Lisa. "How to Write a Theme Statement." *Salem High School. High School.* 21 July 12. http://shslboyd.pbworks.com/f/How+to+Write+a+Theme+Statement+PPT.pdf. Accessed 24Aug10.

Catalina, Jimena. "Reading is Magical. Free PowerPoint Template & Google Slides Theme." , 2021. https://www.slidescarnival.com/brabantio-free-presentation-template/12656. Accessed 30 Oct. 2021.

Cooke, Rachel. "Minds On, Action, Consolidate TCDSB October 2013." Prezi, 31 Oct. 2013. https://prezi.com/nhzxi97179yc/minds-on-action-consolidate-tcdsb-october-2013/. Accessed 30 Oct. 2021.

"Exit Slip Templates." Massachusetts Teachers Association, 2021. https://massteacher.org/-/media/massteacher/files/employment-licensure/ed-evaluation/ddms/editable-exit-slip-templates.pdf?la=en. Accessed 30 Oct. 2021.

"Exit Slips." Reading Rockets, 2021. https://www.readingrockets.org/strategies/exit_slips. Accessed 30 Oct. 2021.

Flowertiare. "Immigrations and passport control at the airport." Under DepositPhotos Standard License.

Fuglei, Monica. "Fun With Words: Boost Reading Engagement With Word Clouds." Resilient Educator, 2021. https://resilienteducator.com/classroom-resources/word-clouds-reading-engagement/. Accessed 30 Oct. 2021.

"Gaining Understanding on What Your Students Know." Edutopia, 23 June 2015. https://www.edutopia.org/practice/exit-tickets-checking-understanding. Accessed 30 Oct. 2021.

Gulley, Joyce and Jeff Thomas. "Using Web 2.0 Tools to Engage Learners." *Star*, 2021.

https://www.collegestar.org/modules/using-web-2-0-tools-to-engage-learners. Accessed 28 Oct. 2021.

Harrison, Don. "Michigan Native American Indian Roadside Chippewas-Odawa-Ottawa Early Village." *Flickr*, under Creative Commons License.

"How to Build a Digital Escape Room Using Google Forms." *Bespoke Classroom*, 2020. https://www.bespokeclassroom.com/blog/2019/10/4/how-to-build-a-digital-] escape-room-using-google-forms. Accessed 30 Oct. 2021.

Jackson, William Henry. "Native American with Medal of President Garfield." *Wikimedia Commons*, Gilman Collection, Gift of The Howard Gilman Foundation, 2005 under Creative Commons License, 2005. https://commons wikimedia.org/wiki/File:Detail_-Native_American_with_a_Medal_of_President_Garfield-_MET_DP275757_(cropped).jpg. Accessed 6 Nov. 2021.

James, Usha. *The Critical Thinking Consortium (TC2)*, 2021. https://tc2.ca/. Accessed 30 Oct. 2021.

Lipika. "What is Web 2.0?" *ZNetLive*, 13 May 2016. https://www.znetlive.com/blog/web-2-0/. Accessed 28 Oct. 2021.

Markusszy. " File:MSzy 20140620 Queen-Abdication-Help.jpg". *Wikimedia Commons*, 20 June 2014. Under Creative Commons License CC BY-SA 4.0, https://commons.wikimedia.org/w/index.php?curid=77550258. Accessed 28 Oct. 2021.

Miller, Matt. "25 Practical Ways to use Google Forms in class, school." *Ditch That Textbook*, 8 Sept. 2019. https://ditchthattextbook.com/20-practical-ways-to-use-google-forms-in-class-school/. Accessed 30 Oct. 2021.

Morton, Donald. "Types of Conflict Worksheet 1." *Ereading Worksheets*, 2020. https://www.ereadingworksheets.com/worksheets/reading/conflict/types-of-conflict-worksheet-01/. Accessed 30 Oct. 2021.

Nedigner, Midori. "What is an Infographic? Examples, Templates & Design Tips." Venngage, 19 Oct. 2019. https://venngage.com/blog/what-is-an-infographic/. Accessed 30 Oct. 2021.

Orooj, Koorosh. "Girl with a Pearl Earring, Johannes Vermeer." Wikimedia Commons under Creative Commons License, 6 Apr. 2016. https://commons.wikimedia.org/wiki/File:Girl with_a_Pearl_Earring(Cropped).jpg. Accessed 6 Nov. 2021.

Ozcinar, Zehra, Regina G. Sakhieva, Elena L. Pozharskaya, Olga V. Popova, Mariya V. Melnik, and Valentin V. Matvienko. Student's Perception of Web 2.0 Tools and Educational Applications. International Journal of Emerging Technologies in Learning (iJET), vol. 15, no. 2, 23 Nov. 2020. https://doi.org/10.3991/ijet.v15i23.19065. Accessed 30 Oct. 2021.

"Question: In the Novel 'The Marrow Thieves': Why is it voice and language and song that destroys the machine?" Chegg, 2021. https://www.chegg.com/homework-help/questions-and-answers/novel-marrow-thieves-voice-language-song-destroy-machine-please-sure-also-provide-quoted-e-q70915992. Accessed 28 Oct. 2021.

Rahimi, Ebrahim, Jan van den Berg, and Wim Veen. "A Pedagogy-driven Framework for Integrating Web 2.0 tools into Educational Practices and Building Personal Learning Environments." The Journal of Literacy and Technology: an International Online Academic Journal, 2021. http://www.literacyandtechnology.org/uploads/1/3/6/8/136889/er.pdf. Accessed 28 Oct. 2021.

Randall, Abby, Amy Schell, and Adriana Romero. "Exit Ticket." The Teacher Toolkit, 2021. https://www.theteachertoolkit.com/index.php/tool/exit-ticket. Accessed 30 Oct. 2021.

Rolling Fishays. "Steel Rails Fall." Under Depositphotos Standard License.

Rowe, Erin and Lesley Chapel. "Web 2.0 Tools for Education." Study, 2021. https://study.com/academy/lesson/web-20-tools-for-education.html. Accessed 28 Oct. 2021.

Rubens, Peter Paul. "Portrait of a Franciscan Friar." Wikimedia Commons under Creative Commons License, 10 Aug. 2017. https://commons.wikimedia.org/wiki/File:Rubens_Portrait_of_a_Franciscan_friar_01.jpg. Accessed 6 Nov. 2021.

Salazar, Ray. "If You Teach or Write 5-Paragraph Essays–Stop It!" The White Rhino: A Chicago Latino English Teacher, 10 May 2012. http://www.chicagonow.com/white-rhino/2012/05/if-you-teach-or-write-5-paragraph-essays-stop-it/. Accessed 8 July 2012.

Searagan. "Teepee Village." Under DepositPhotos Standard License.

Selwyn, Neil. "Web 2.0 applications as alternative environments for informal learning—a critical review." Institute of Education, University of London, UK, 2017. http://newinbre.hpcf.upr.edu/wp-content/uploads/2017/02/39458556-W2-informal-learning.pdf. Accessed 28 Oct. 2021.

Stannard, Russell. "How to use TED-ED—Ideas for the Flipped Classroom." YouTube, uploaded by Russell Stannard, 30 May 2018. https://www.youtube.com/watch?app=desktop&v=1fL9YIxMB88&feature=youtu.be. Accessed 30 Oct. 2021.

Tammy. "24 Exit Ticket Ideas." The Owl Teacher, 2021. https://theowlteacher.com/24-exit-ticket-ideas/. Accessed 30 Oct. 2021.

Tatli, Zeynep, Hava İpek Akbulut, and Derya Altınışık. "Changing Attitudes Towards

Educational Technology Usage in Classroom: Web 2.0 Tools." *Malaysian Online Journal of Educational Technology*, vol. 7, no. 2. https://files.eric.ed.gov/fulltext/EJ1214029.pdf. Accessed 28 Oct. 2021.

Schrock, Kathy. "The Critical Thinking Workbook." *Kathy Schrock's Guide to Everything*, 2021. https://www.schrockguide.net/uploads/3/9/2/2/392267/critical-thinking-workbook.pdf. Accessed 5 Nov. 2021.

SharryX. "File: Rock-paper-scissors Shapes.png." *Wikimedia Commons,* under Creative Commons Attribution-Share Alike 4.0 International license, 7 Aug. 2020. https://commons.wikimedia.org/wiki/File:Rock-paper-scissors_Shapes.png. Accessed 6 Nov. 2021.

The Ontario Curriculum Grades 9 and 10: English. Ontario Ministry of Education, 2007. http://www.edu.gov.on.ca/eng/curriculum/secondary/english910currb.pdf. Accessed 28 Oct. 2021.

Wakeford, Larry. "Sample Exit Tickets." Brown University: The Harriet W. Sheridan Center for Teaching and Learning, 2021. https://www.brown.edu/sheridan/teaching-learning-resources/teaching-resources/course-design/classroom-assessment/entrance-and-exit/sample. Accessed 30 Oct. 2021.

"What is an Infographic?" Infogram, 2021. https://infogram.com/page/infographic. Accessed 30 Oct. 2021.

Zappa, Joe. "The Possibilities of Web 2.0 and Google Apps." Classroom 2.0, 15 June 2015. https://classroom20.com/profiles/blogs/the-possibilities-of-web-2-0-and-google-apps. Accessed 28 Oct. 2021.

www.ingramcontent.com/pod-product-compliance
Lightning Source LLC
Chambersburg PA
CBHW060859270326
41935CB00003B/28